YOUR KNOWLEDGE HAS VALUE

- We will publish your bachelor's and master's thesis, essays and papers

- Your own eBook and book - sold worldwide in all relevant shops

- Earn money with each sale

Upload your text at www.GRIN.com and publish for free

John Tarilanyo Afa

The Effect of Environmental Factors on Electrical Machines

GRIN Verlag

Bibliografische Information der Deutschen Nationalbibliothek:

Die Deutsche Bibliothek verzeichnet diese Publikation in der Deutschen National-
bibliografie; detaillierte bibliografische Daten sind im Internet über http://dnb.d-
nb.de/ abrufbar.

Imprint:

Copyright © 2011 GRIN Verlag GmbH
Druck und Bindung: Books on Demand GmbH, Norderstedt Germany
ISBN: 978-3-656-41234-2

JOHN TARILANYO AFA

CURRICULUM DESIGN

TECHNICAL REPORT

FEM 633:

EFFECT OF ENVIRONMENTAL FACTORS ON ELECTRICAL MACHINES

ATLANTIC INTERNATIONAL UNIVERSITY

HONOLULU, HAWAII

2011

TABLE OF CONTENT

Section   **Page**

1.0   Introduction --- 1

2.0   General Analysis --- 4

       2.1 Basic Laws of Electromagnetism --- 4

       2.2 Forces and Torques in Magnetic field system --------------------------- 4

       2.3 Insulations of Machine -- 6

3.0   Materials and Methods -- 9

4.0   Results --- 10

5.0   Discussion -- 11

6.0   Recommendation --- 15

7.0   Conclusion --- 16

8.0   References --- 17

FAULTS ON ELECTRICAL MACHINES AND THE ENVIROMENT

1.0 INTRODUCTION:

A machine is a device that does useful work in a predictable way according to physical Laws. A machine could be referred to a motor or a generator and in most cases is reversible, therefore electro mechanical energy conversion devices are necessary. Motor and generator which are in the category of continuous energy conversion equipment in an electro mechanical energy conversion uses magnetic medium. With magnetic field more than a million joules can be stored in a cubic meter of air but with an electrostatic medium breakdown of the air limit the storage to few joules. The motoring and generating action is basically a reversible process that is, the same machine can be used as a generator or a motor and shown in fig. 1

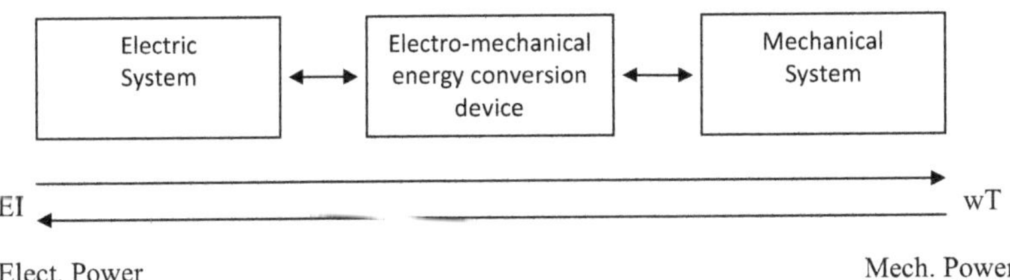

Fig. 1: Reversible Process of Energy conversion.

Electro-mechanical energy conversion device is a link between an electrical and mechanical system and electromagnetic system needs the presence of natural phenomena which inter-relate electric and magnetic fields on one hand and mechanical force and motion on the other hand.

2.0 GENERAL ANALYSIS

2.1 Basic Laws (Electromagnetic)

The two different processes of emf development can be termed the motion or speed emf and the transformer emf. Although in the final analysis they are the same.

If a conductor of length 1 meters is moved with a velocity of V m/s in a magnetic field of constant flux density B teslas (w|A|m^2), then the emf induced in the conductor is equal to the rate of flux linkage and is given by

$$E = Blv. \qquad \dots \dots \dots \dots \dots \dots \dots \dots \dots \dots \quad (1)$$

The transformer emf relates to the emf that is set up in a coil when the magnetic flux linking the coil is varying with time and is given by

$$E = \text{Rate of change of flux linkage}$$

$$= \frac{d\varphi}{dt} = - N\frac{d\phi}{dt}$$

Where φ is the flux linkage of the coil and is equal to the product of the number of turns of the coil N, and flux ϕ in Weber, linking the coil.

2.2 Forces and Torques in Magnetic Field Systems

According to Lorentz force Equation

$$f = q\left(E + V \times B\right). \qquad \dots \dots \dots \dots \dots \dots \dots \dots \dots \quad (2)$$

Where f is the force in Newton on a particle of charge q coulombs in presence of Electric and Magnetic fields, E is volt per meter, B in Tesla and V is the velocity of the particle relative to the magnetic field in m|s.

In magnetic field system the force is given as

$$f = q(v + B) \ldots \ldots \ldots \ldots \ldots \ldots (3)$$

That is, it is determined by the magnitude of the charge on the particles and the magnitude of flux density B as well the velocity of the particles. The direction of force is always perpendicular to both the direction of particle motion and the direction of magnetic field.

For situations where large numbers of charge particles are in motion it is convenient to write the equation (3) in terms of current density J, which is

$$f = J \times B \; N\,|\,m^3 \ldots \ldots \ldots \ldots \ldots (4)$$

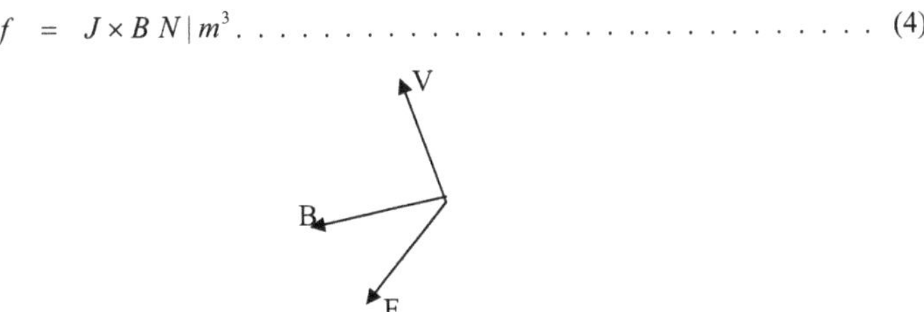

Fig 2: Force, Motion and Magnetic Field direction

The mass of the materials used in the construction of electrical machines remains constant under the conditions of operation, therefore the principle of conservation of energy can be applied in the energy conversion.

The input energy must therefore be equal to the summation of the useful output energy, the energy converted into heat and the change in the energy stored in the magnetic field. The energy balance equation may be written as follow:

$$\begin{pmatrix} \text{Electric energy input} \\ \text{less resistance or} \\ \text{olimic loss} \end{pmatrix} = \begin{pmatrix} \text{Mechanical energy} \\ \text{output plus friction and} \\ \text{windage losses} \end{pmatrix} + \begin{pmatrix} \text{Increase in energy stored} \\ \text{in coupling field associated} \\ \text{losses} \end{pmatrix} \ldots (5)$$

This can be diagrammatically be represented as shown in fig. 3

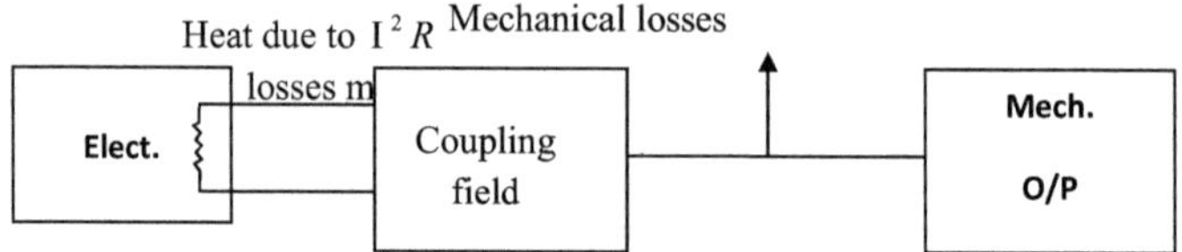

Fig. 3: Electric Mechanical energy conversion Device.

The left hand side of equation (5) can be express in terms of current and voltages as

$$dW_{elect} = eidt = vidt - i^2rdt$$

Generally, the equation (5) can be represented as

$$dW_{elect} = eidt = dW_{field} + dW_{mech}$$

Where dW_{mech} is the differential energy converted into mech form dW_{field} is the differential energy absorbed by coupling field.

2.3   INSULATIONS OF MACHINES:

The winding of each phase of a Rotating machine consist of coil connected in phase and placed in slots in an identical manner. The number of turns in each coil depends on the nominal voltage and the power output of the machine and also on the speed of rotation of its rotor (Farahani, et al 2005, Tom Bishop 2008). In large capacity machines parallel conductors are used for the reduction of eddy current loss.

(iii)   Insulation between coil sides

Insulation in rotating machines is continuously subjected to vibration and impact mechanical load in operation, therefore it must possess high mechanical strength and monolithicnes (Deshphande, et al, 2011.Njmbola, et al 2008, Contin, et al 2006).

In electric machines different insulating materials are used in the same machines because of the following:

(i)   Insulations between turns of a coil

(ii)   Insulations relative to the frame

(iii)   Insulations between parallel conductors of a turn

(iv)   Insulations between coil sides that are between the sides of different coils placed in one and the same slot.

The insulating materials begin to degenerate at relatively small temperature. For reliable operation, it is, essential that the temperature rise in electrical machines do not exceed the permissible maximum temperature of the insulating materials used therein. The three fundamental electrical properties of materials of utmost importance in the operation of electrical machines are insulation resistance, dielectric strength, and dielectric loss angle. Others are mechanical strength heat resistance, hygroscopicity and it must be capable of withstanding a repeated heat cycle without deterioration.

Several materials are used as insulating materials and are grouped into different classes.

Dielectric characteristics of insulation materials of rotating system can be affected by operational and environmental conditions. Humidity, temperature and contaminations affect dielectric parameters such as insulation resistance, polarization index and dissipation factor in certain degree. In large machines the effect of partial discharge due to some recognize factors including aging, has become an area of interest.

Partial discharges are small electric sparks that occur within the electric insulation of switchgear, cables, transformers and windings of large motors and generators (John Wilson 2004, Candela, et

al 2000). Just as every material has a characteristic tensile strength, each material also has an electric breakdown (dielectric) strength that represents the electric intensity necessary for current to flow and an electric discharge to take place. Common insulating materials such as epoxy, polyester and polyethylene have very high dielectric strengths. Conversely air has a relatively low dielectric strength. Electric breakdown in air causes an extremely brief electric current to flow through the air pocket. The measurement of partial discharge is the measurement of these breakdown currents (Vogelsang, et al 2005, Birlasekan, and Xmgzhou, 2003)

Electric equipment can suffer from a manufacturing defects or operational problems that impair its mechanical reliability but the external conditions (environmental) has great role to play in the reliable operation of motors and generators. In all cases these stresses cause as a result, will weaken the bonding properties of the insulating materials of the windings. Not only do partial discharge levels provide early warning of imminent equipment failure, but partial discharge also accelerates the breakdown process.

Heat is developed in all electrical machines due to losses in the various parts of the machine. The principal losses are due to copper losses (I^2R loss), eddy current and hysteresis loss in iron, mechanical losses due to friction in bearing, air friction or windage losses. The temperature rise depends on (i) the amount of heat produced (ii) amount of heated dissipated per I^0 c rise of the surface of a machine.

According to Newton law of cooling the rate of loss of energy of a hot body is proportional to the difference in temperature between that body and its surroundings. Heat balance equation can be put in the following:

Energy converted into heat = Heat absorbed + Heat dissipated

$$P dt = mC_r d\theta + S\theta \times dt \cdots\cdots\cdots\cdots\cdots\cdots (7)$$

Due to this energy balance the type of enclosure, the duty cycle, and the environmental factors are of great importance in the selection of the machine type.

3.0 MATERIALS AND METHODS

The factors were taken in recognition of these three categories of rotating machines:

(i)   The fractional horse power motor machines

(ii)   The low voltage rotating machines

(iii)   The high voltage rotating machines

The effect of atmospheric over voltage (lightening) on the high voltage machines is very significant

For this reasons the Humidity, Rainfall, Temperature, wind and dust particles, salt spray and atmospheric over voltages were considered. These results were compiled from two different stations and are complimented by the giving literatures.

Sixteen different sites were chosen and the average annual records of rainfall, humidity, lightning days and temperature were taken. These values were taken for the two seasons (dry and rainy seasons) in the year.

4.0 RESULTS:

The results are shown in table 1

Atlantic International University

Table 1: Relative humidity, Annual Rainfall, Temperature and Lightning days

S/N	Place/Location	Season	Relative Humidity	Mean Annual Rainfall	Main Annual Temp. (^{0}C)	Lightning Days
1.	OPOBO	W	90-95	3816.8	26	90
		D	61-65		31	
2.	AHOADA	W	95-85	2370.1	25	80
		D	53-56		31.2	
3.	ONNEY	W	80-85	2425	27	80
		D	55-60		30	
4.	BORI	W	75-80	2287.5	22	65
		D	52-55		31.5	
5.	DEGEMA	W	80-90	2355.1	25	75
		D	55-55		30.5	
6.	ONNE	W	85-90	2438.4	26.2	75
		D	60-65		31	
7.	PORT HARCOURT	W	75-78	2370.5	28	75
		D	48-54		32	
8.	YENAGOA	W	80-89	2484.2	24	78
		D	50-53		31.2	
9.	BRASS	W	84-89	2512.2	25	85
		D	54-62		29	
10.	NEMBE	W	85-92	2705.2	25.2	87
		D	59-63		31.2	
11.	SAGBAMA	W	87-91	2452.7	26	85
		D	56-62		30.7	
12.	AKASSA	W	89-93	2951.3	24.8	90
		D	61-65		30	
13.	OPOROMA	W	80-87	2492.5	26	75
		D	60-64		31	
14.	OLOIBIRI	W	80-90	2354.6	27	69
		D	56-61		32	
15.	WARRI	W	19-84	2289.1	26.5	70
		D	50-55		31.6	
16.	KULA	W	90-94	2932.7	24	120
		D	65-70		30.1	

W = Wet (Rainy Season)
D = Dry (Dry Season)

5.0 DISCUSSION:

The rainfall in the area is seasonal. The rainy seasons starts from March and increase to July and breakup at the middle of August (August Break), and have the heaviest rainfall by September and gradually reduces to the dry seasons in November to February. The seasonal variations affect the humidity level.

From the table (Table 1) the average relative humidity can be taken from 52% during the dry seasons and 95 percent during rainy seasons. From the sampled areas the lowest humidity level is above the upper limit of the critical relative humidity level. Therefore at favorable conditions condensation can take place most of the year except during harmattan period.

The tested results are shown in table 2.

Symptom	Renst meas	Likely cause	Phases	Usuage	No
Arm. Hot all over	Low resistance	Dampness	1ph	Grinding	4
Burnt out winding	Low resistance	Dampness	1ph	Water pumping	5
Short at in fields	Low resistance	Dampness	1ph		4
Sparking at brush	Very low winding resistance	Brush not at neutral point	1ph	Grinding	2
Partial short circuit in field windings	Very low resistance	Moisture in coil	1ph	Compressor motor	5
Burn in one winding	One winding burnt out others below normal	Fault on winding	3Φ	Running Compressor	2
Not running well (hot)	Low resistivity for all phases	Moisture in windings	3Φ	Running pump	3
Hot and smoking at windings	All winding low	Moisture on winding	3Φ	Lathe motor	4
Hot and low output	Winding resistivity below normal	Dry Bearing/Moisture	3Φ	Lift motor	2

From the results in table 2, moisture is a major cause of deterioration of winding insulations, burning etc.

When any warm dry motor is shut down, it draws in fresh cool air which always contains moisture even in hot dry weather. After the motor has cooled, this moisture condenses within the porous winding insulation. On humid days or during rainy weather, the air contains much more than the normal amount of moisture and the insulation absorbs more.

If the motor is operated continuously, or frequently, the insulation remains sufficiently dried out to prevent insulation failure. If the motor stands idle for a long periods, especially in unheated building during humid weather the trapped moisture may accumulate until the insulation becomes a partial conductor of electricity. If normal voltage is then applied, the insulation material may fail, either by grounding or phase to phase fault for three phase machine.

Partial discharge: For large machines the partial discharge phenomenon is a regular occurrence. Dielectric characteristics of insulation for such rotating machines can be affected by operational and environmental conditions. In a humid condition like the Niger Delta area, partial discharge activities or coronation in machines will be frequent.

Partial discharges which are a locally confined electrical breakdown within the high voltage insulating system may occur internal to insulation or on the surface or a sharp point. Several factors are responsible for partial discharge in a H.V rotating machines including aging of the insulation but humidity is a major factor. However, there are no formulae or "rules of thumb" for the effect of humidity with respect to rotating electrical machines windings [Neta world]. Though partial discharges is localized but may cause progressive deterioration of the insulation which may eventually lead to breakdown.

Dust particles: Dust particles affect insulations of windings, bearings and other rotating parts of the machines. In the area of this study such harmful accumulations may not take place within two years if normal scheduled maintenance is carried out. The only period the effect of dust is felt is the harmattan period which is brought about by the North East wind from the Sahara which last for about 4 to 6 weeks.

The marine environment: The effect of marine environment is more sever and about two third of the land is affected by this environment. The sea air contains chlorides and may contain traces of sulphur, carbon dioxide and other components that over a period of time can become concentrated on metal surfaces. Furthermore, in areas immediately adjacent to the shore, the salt spray from the ocean thrown up by strong breezes contributes to the build up of sea salt deposits on metal which keeps them wet due to the high humidity in the area. These effects may be on casing of rotating machines, bearings and all metal parts including windings. If the machine is kept from operation for a long period of time the insulations may also become strong and bristle with salt bearing and rust on parts of machines.

Temperature effect: The temperature rise in machines may be due to faulty bearing, overload and high Ambient Temperature. The temperature recorded in the area is about 23^0c to 32^0c. The maximum temperature range for small machine is about 70^0c to 120^0c for H.V (big Hp) motors. Since the difference in temperature range is high, energy dissipation and cooling may not be the problem, that is, when temperature alone is taken in isolation (as an external factor).

Lightning effect on h.v machines: The average lightning days for the year is between 70 to 100 days in the coastal areas, therefore its effect on rotating H.V machine is very pronounced. Some rotating machines are connected directly to overhead lines and are exposed to lightning surges.

Other H.V electric machines are connected through transformers and are exposed to travelling waves by lightning strikes. Oil is used on circuit breakers and transformers as an insulation to smooth the stress that results from electric voltages. In case of rotating machines there is no oil insulation and the windings are equipped with dry insulation. Therefore rotating machines are more venerable to high stress failure as compared to circuit breakers or transformers. With the fast front transients, the winding structure can cause surge reflections and oscillations that can damage winding insulations. Therefore H.V rotating machines are exposed to travelling wave, switching surges as well as other internal over voltages. It is therefore necessary that H.V rotating machines are protected as substations.

6.0   RECOMMENDATIONS:

❖   It is a good practice to record the humidity reading as well as winding temperature and insulation resistance for each insulation resistance test. Knowing these parameters, insulation resistance readings can be evaluated for possible effect of humidity, recognizing that at high humidity the insulation resistance reading may be lower for a specific temperature.

❖   Special provision should be made to protect motor windings which operate in a damp atmosphere or are subject to condensation of moisture when out of service for long periods.

Where electric power is not available or cannot be economically justified to keep motor warm during long shutdown periods, the most practical solution is to provide moisture proof covers of plastics or water proof fabric.

These covers should be bound tightly around the shaft and motor base so as to be as tight as possible. Some reduction in moisture may be obtained by placing a drying agent such as silica gel in porous bag inside the cover (Stone, et al 2004, Mousa, et al 2008).

❖ A standard schedule should be worked out which provides for measuring the insulation resistance of each motor at least once a year. In addition, those motors which have been idle for long periods, especially those in humid atmosphere should be measured before starting.

❖ Whether online or off-line partial discharge (PD) test should be done at regular interval (at least once a year) on big machines to know the integrity of the insulation material.

7.0 CONCLUSION:

From the available records the main problems in the small motors is the moisture content in the air especially those machines that are operated in damp areas. From statistics it was seen that those machines that are used for water pumping undergoes repairs atmost every six months.

Another problem is rusting on the metal bodies. When a machine is not operated for six months to one year most rotating part become stiff and shows sign of rust in some areas. This is due to the coastal effect and the level of humidity in the environment

For big machines that are either connected directly to lines or through transformer and adequate protection from lightning is necessary for safe operation. Lubrication and partial discharge testing of machines is necessary at least once a year to give a longer life cycle of the machine.

8.0 REFERENCES:

1. Deshphande, A.S., Patil, A.S., Cheeran, A.N., 2011: The Application of Weibull Function to Partial Discharge Analysis and Insulation Agency: A Review. International Journal of Advanced Engineering Science and Technologies 3 (2), 111 – 114.

2. John Wilson, (2004). Partial Discharge Analysis, Ultrasonic Technologies to Evaluate Partial Discharge in Electrical Machinery SKF Reliability Systems U.S.A www.aptitudexchange.com

3. Farahani, M., Borsi, H., Gockenbach, E., 2005. Dielectric Response Studies on Insulating System of High Voltage Rotating Machines 20th International Power System Conference 98 – E – WL. M – 443.

4. Tom Bishop, P.E., 2008. Insulation Resistance Testing: How Many Megohms does it take to start a motor. Electrical Apparatus Service Association (Spring 2008, NETA WORLD), www.netaworld.org

5. Vogelsang, R., Fruth, B., Frohlich, K., 2005. Detection of Electrical Tree propagation by Partial Discharge measurements. European Transactions on Electrical Power ETEP 15, 1 – 14.

6. Nimbole, V., Lakdawale, V., Basappa, P., 2008. Prediction of Partial Discharge Pulse Height Distribution Parameters using Linear Prediction Method Conference on Electrical Insulation Dielectric Phenomena. Pp 337 – 340.

7. Candela, R., Mirelli, G., Schifani, R., 2000. PD Recognition by means of Statistical and Fractal Parameters and a Neural Nature IEEE Transaction of Elect. Insulation (7): 87 – 94.

8. Contin, A., Cavallini, A., Montanari, G.C., Hudon, C., Belec, M., Ngeyen, D.N., 2006: Searching for Indixes Suitable for Rotating Machines Diagnosis. IEEE International Symposium on Electrical Insulation. Pp 101 – 105.

9. Nedjar, M., Beroual, A., Ageing under A.C Voltage of Polynethene by using weibull statistic, 2004. International Conferences on Solid Dielectric U.K. pp. 654 – 657.

10. Secong-Hee Park, Hae-Eun Jung, Jae-Hun Yun, Byoung-Chul Kim, Secong-Hwa Kang, Kee-Joe Lim 2008. Classification of Defects and Evaluation of Electrical Tree Degradation in cable Insulation Using Pattern Recognition Method and Weibul Process of Partial Discharge International Conference on condition monitoring and diagnosis China.

11. Mousa VI Gargari, Wouters, S., Wielen, P.A.A.F, Vander, P.C.J.M., Steenrus, E. 2008. Proc. International Conference on Probabilistic Methods Applied to Power Systems Puerto Rico.

12. Stone, G.C., Boulter, E.A., Culbert, I., Dhirani, H. 2004. Electrical Insulation for Rotating Machines-Design, Evaluation, Aging, Tesing and Repair. IEEE Press. ISBN 0-471-44506-1.

13. Zaengi, W.S., 2003. Dielectric Spectroscopy in Time and Frequency Domain for H.V. Power Equipment, Part 1: Theoretical Consideration. IEEE Electrical Insulation Magazine 19 (5) 5 – 19.

14. Kauthold, M., Schaefer, K., Baucer, K., Bethge, A., RIsse, J., Interface Phenomena in Stator Winding Insulation – Challenges in Design, Diagnosis and Service Experience 18 (2), 27 – 36.

15. Birlasekan, S., Xmgzhou, Y., 2003. Relaxation Studies on Power Equipment. IEEE Transaction on Dielectrics and Electrical Insulation 10 (2) 1061 – 1077.